Daniel Chitata
Alfredo Muacahila
Sebastião Tumitângua

Living Forest, Strong Community

Daniel Chitata
Alfredo Muacahila
Sebastião Tumitângua

Living Forest, Strong Community

**Contribution to sustainable forest management in the
Munhino Community, Namibe Province**

ScienciaScripts

Imprint

Any brand names and product names mentioned in this book are subject to trademark, brand or patent protection and are trademarks or registered trademarks of their respective holders. The use of brand names, product names, common names, trade names, product descriptions etc. even without a particular marking in this work is in no way to be construed to mean that such names may be regarded as unrestricted in respect of trademark and brand protection legislation and could thus be used by anyone.

Cover image: www.ingimage.com

This book is a translation from the original published under ISBN 978-620-6-76033-7.

Publisher:
Sciencia Scripts
is a trademark of
Dodo Books Indian Ocean Ltd. and OmniScriptum S.R.L publishing group

120 High Road, East Finchley, London, N2 9ED, United Kingdom
Str. Armeneasca 28/1, office 1, Chisinau MD-2012, Republic of Moldova, Europe
Printed at: see last page
ISBN: 978-620-7-62241-2

Dedication

To God for the gift of life. To my dear parents Domingos Chitata and Joana Chiteculo, my wife Florence Cassoma E. Chitata, my siblings Luciano Chitata, Ana Chitata, Delfina Chitata, my colleagues, friends, for having been the driving force and tireless encourager in the completion of my endeavor and to all those who directly or indirectly contributed to the success of my academic training.

Thanks

At the end of this course, I can only thank God, the Creator and Almighty, for having enlightened me and guided me along the paths that have led me here, for the safekeeping, health and victory that have contributed to this achievement.

To my dear parents Domingos Chitata and Joana Chiteculo and my siblings, for their encouragement, support and the patience with which they have always helped me.

To my family, especially my wife Florence Cassoma Ernesto Chitata, for the patience, insight and kindness with which they put up with my absences at times reserved for them.

From the bottom of my heart, I would like to express my sincere and profound thanks to my supervisor, Professor Alfredo Noré Muacahila, for accepting the supervision of this dissertation, for his support, his valuable suggestions for the work, for understanding my difficulties as a student and, above all, thank you, thank you for accompanying me in this difficult task and motivating my interest in knowledge and academic life, for your wise guidance in the most complicated moments when the light seemed to be nowhere to be seen at the end of the tunnel.

I would like to thank my parents Paulino Kapewa and Domingas Kapewa and my siblings Edith, Cândy, Neide Kapewa, Njundo, Gegé, Nairo and Nguevinha. Thank you from the bottom of my heart for not letting me lack bread and the warmth of family life, which was crucial to my physical and mental health during my formative years in Lubango.

I would also like to thank my fellow students, the teaching staff and the coordinators of the Master's course in Ecology and Natural Resources Management, 2018/2020 edition, for the excellence of the training provided and the knowledge imparted, which was very useful in completing this work.

I would like to thank the staff of the Forestry Development Institute in Moçâmedes and all the respondents for their availability.

Of course, the list is enormous, so I would like to thank everyone who directly or indirectly contributed to the success of this work and my education in general.

Thank you very much!

Summary

Given the current environmental crisis, one of the reasons for which is the unsustainable way in which natural resources are exploited, actions and contributions that contribute to sustainable development are necessary. The aim of this study was to analyze the causes of forest degradation in the community of Munhino, in the municipality of Bibala, and, based on scientific and local knowledge, to contribute to sustainable forest management, taking into account the local social and economic reality. The methodological procedure thus included interview and questionnaire surveys, direct observation of the situation, interaction with the local population, as well as a literature review on the subject (legislation, reports/information from local authorities, academic works, magazines and documents from international organizations). The data collected was statistically processed using IBM SPSS Statistics 22 software and Excell software from the Microsoft Office student 2013 package. From the analysis and discussion of the results, it was possible to identify direct and indirect factors that contribute to the unsustainable management of the local forest, above all the lack of job opportunities and poverty (as determining factors), seconded by a set of underlying factors such as: misaligned public actions, poor supervision and the absence of socio-environmental education.

Keywords: Forests, Sustainable management, Munhino, Poverty.

PREFACE

Dear Reader,

It is with great pleasure that I present the literary work entitled "Living Forest, Strong Community: a contribution to Sustainable Forest Management in the Munhino Communities". This book is more than just a collection of practices and guidelines; it is a testimony to the power of community unity and commitment to environmental preservation and sustainable development.

This book was inspired by the findings and experiences of the residents of Munhino themselves, as in recent years we have faced growing challenges due to the unsustainable exploitation of natural resources and, as a consequence, the climate changes that are devastating Planet Earth. It was in this context that the urgent need arose to contribute to the mitigation of local environmental problems by drawing up strategies that were found to be relevant to ensuring that our forests can continue to thrive and sustain future generations by guaranteeing a thermal balance at local level and beyond.

In this sense, the idea that the rational implementation of forest exploitation activities by communities should be based on the promotion of environmental awareness and education actions, in the sense of saving today to have tomorrow, and the exercise of environmental education should be built and enriched both with the theoretical and scientific foundations of good practices in the exploitation of forest resources, as well as with local socio-economic experiences.

During the preparation of this work, we had the collaboration of various experts and community leaders and I would like to express my deep gratitude to all those who contributed their knowledge and experience, especially the families who welcomed us and shared their traditional practices. This text "Living Forest, Strong Community" is a call to action. We hope that each page will inspire you to become actively involved in sustainable forest management, promoting practices that benefit both the environment and the community, and that it will be a powerful tool in building a future where forests and community grow together, strong and vibrant.

Before concluding, I would like to thank the organizers of the book, whom I congratulate on their initiative which, I believe, could make a significant contribution to improving the quality of future research into environmental issues in Angola and other parts of the world.

INTRODUCTION

I NTRODUCTION

In order to improve his social and economic living conditions, man has sought to exploit natural resources to the point of exhaustion without any concern for sustainability, forgetting the laws that govern nature to ensure environmental balance.

This short-term materialistic attitude or vision is common in both industrialized and developing countries, albeit with different dynamics and levels of development. While in the first case, the alteration of the environment is due to high levels of production and consumption, in the second, the degradation of natural resources is due, in general, to the extreme poverty to which the majority of populations are exposed and which rely on practices such as deforestation, thus devastating landscapes and their *habitats*. Thus, to a lesser or greater extent, both countries are responsible for the current global environmental crisis, which is aggravated by the emission of greenhouse gases and other gases that contribute to environmental pollution, rising global temperatures, the appearance of acid rain, desertification and a decline in biodiversity.

In the municipality of Bibala, the environmental problem, especially the issue of exploitation of forest resources, is a reality that inspires great concern, as the population seems to have no idea of how important it is to guarantee the sustainability of this resource, the forest. It seems that the population is only concerned with guaranteeing their daily survival without considering the forest's resilience in revitalizing itself. In this part of the country, deforestation has been recurrent and has become almost the *modus vivendi of* the local communities, subjecting the terrestrial ecosystems present there to a situation of accentuated degradation.

This book is organized into six parts: the Introduction - in which the problem under study is highlighted, the importance of its study, the objectives pursued, as well as the methodological design supporting the research carried out; Chapter I - in which the author describes through literary concepts on the subject under approach, giving an account of some basic concepts on issues related to the exploitation of natural resources, in particular forests, the state of forests worldwide and Angola in particular, aspects of sustainable forest management with a view to eliminating practices that lead to deforestation of forests, among other issues; Chapter II - which sets out the methodology used to carry out the study, as well as the characterization and location of the study area; Chapter III - which analyses and discusses the results of the surveys carried out, interacting with the concepts and knowledge mentioned in Chapter I; Chapter IV - where, taking into account the aspects covered in the previous topics and especially the results obtained in Chapter III, we present the actions/activities that we think would be useful, if put into practice, to minimize and control the mismanagement of the local forest that we are witnessing and which leads to its deforestation; and the Conclusion - which highlights the most striking aspects of the study carried out, according to the objectives and the problem raised.

This book is recommended reading for the academic community and the general public, given the relevance of research into environmental issues resulting from anthropogenic actions and their influence on the environment, with the aim of arousing interest in preserving forest areas.

CHAPTER I

FORESTS: ESSENTIAL ECOSYSTEMS FOR LIFE ON EARTH

Chapter1. FORESTS: INDISPENSABLE ECOSYSTEMS FOR LIFE ON EARTH

1.1 General aspects

Before discussing forests and their management, the benefits they bring to life on Earth on the one hand, and the problems these terrestrial ecosystems face on the other, it is important to clarify some key concepts that will help us understand this subject, namely: forest, deforestation, deforestation, forest degradation and other related concepts.

- *Forest according to* (FAO, 2007) is an extension of land of more than 0.5 hectares, with a height of more than 5 meters and a canopy cover of more than 10%, with the exception of land dedicated to agriculture and urban uses. In less technical language, a *forest is* any terrestrial ecosystem with a cover of trees, shrubs or any other type of vegetation, including wild animals and microorganisms. In the Angolan legal system, from a utilitarian point of view, two types of forest are distinguished (DecPre, 171/18): *natural forest* - that which is not the result of human action, but is also immune to it; and *community forest* - that which is located on land classified as community.

- *Deforestation* - is the conversion of forest or part of it to another type of land use (agriculture, grazing, urban construction, ...), or the long-term reduction of its trees to below the 10% threshold (FAO, 2001). Simply put, deforestation is "the indiscriminate destruction or cutting down of trees without proper replacement" (DecPre, 171/18). Associated with the concept of deforestation is the concept of *Deforestation,* which is sometimes confused with it. *Deforestation* therefore represents the way in which deforestation is processed, according to the concept given by (DecPre, 171/18): "clear cutting or total destruction of vegetation in a given area".

- *Sustainable Use* - is the management and exploitation of forest resources without compromising the ecological functions of forests, so as not to harm the ecological, economic, social and landscape value of the ecosystem, which must be at the service of current and future generations (DecPre, 171/18).

- *Forest Resource* - represents the tangible and non-tangible goods provided by the forest, of current or potential value to humanity (living organisms, any product found within the forest, the ecosystem services of forests) (DecPre, 171/18).

- Forest Degradation - is the long-term reduction in the total potential benefits of the forest, including carbon, timber, biodiversity and other goods and services (honey and other foods), caused mainly by direct human activity (FAO, 2006; ITTO, 2005; IPCC, 2003).

1.2 Ecosystems and their relationship with the Biosphere

Ecosystems represent the set of living organisms (biotic component) and the physical environment (abiotic component), plus the interactions between living organisms and between them and the environment itself (Carapeto, 2004). Examples of ecosystems are deserts, natural grasslands, forests, lakes, wetlands, oceans, among other places.

The economic value or cost of the environmental resources provided by ecosystems is often underestimated and not observed on the market through the price system. In general, their economic value comes from their attributes, which may or may not be associated with a use (Table 1) (Motta, 2007; Faucheux and Noel, 1995).

*Table 1 **Economic value of environmental goods and services** (Source: Adap. from Motta, 2007, p. 13).*

Use value			Non-use value
Direct	Indirect	Option	Existence
Environmental goods and services appropriated through the direct exploitation of resources for consumption today.	Environmental goods and services generated by ecosystem services and indirectly appropriated for consumption today.	Environmental goods and services of direct and indirect use to be appropriated in the future.	Value not associated with current or future use, but which reflects moral, cultural, ethical or altruistic issues.

The literature presents theoretical foundations and methods that help, if necessary, to determine the economic value of environmental resources and services (Figure 2). The following expression presented by (Motta, 2007) is one of them:

$$VERA = (VUD + VUI + VO) + VE$$

Where:

DUV (direct use value) - the value that individuals attribute to an environmental resource because they use it directly (in the form of extraction, direct consumption or as a recreational object);

VUI (indirect use value) - the value that individuals attribute to an environmental resource, deriving the benefit of its use from ecosystem functions (in containing erosion, serving as a *habitat for* species reproduction);

VO (option value) - the value attributed by individuals to preserving resources that are under threat, in order to guarantee direct and indirect uses in the near future;

EV (non-use value or existence value) - value that disassociates the use of the resource for cultural, ethical and other reasons from the existence of species other than humans (human actions to protect some natural species).

As ecosystems that provide goods and services, forests are places that play a leading role in the context of life on Earth. They are part of the "reasons for life on Earth" formula[1] , as they provide important services that benefit terrestrial life and consequently the well-being of humans and other forms of life. Figure 1 presents a grouping of the main functions and services provided by ecosystems in general, and by forests in particular, showing that their main function is to support the existence of life on earth (through biogeochemical cycles, soil formation, achieving productivity), thus ensuring various services: food, fresh water, clean air, wood, fibers and fuel, *habitat,* biological and medicinal products, climate regulation (WHO, 2005).

---

[1] This expression embodies the subtitle of Clemente's book (Clemente, 2020), which highlights the role of forests in decarbonizing the atmosphere.

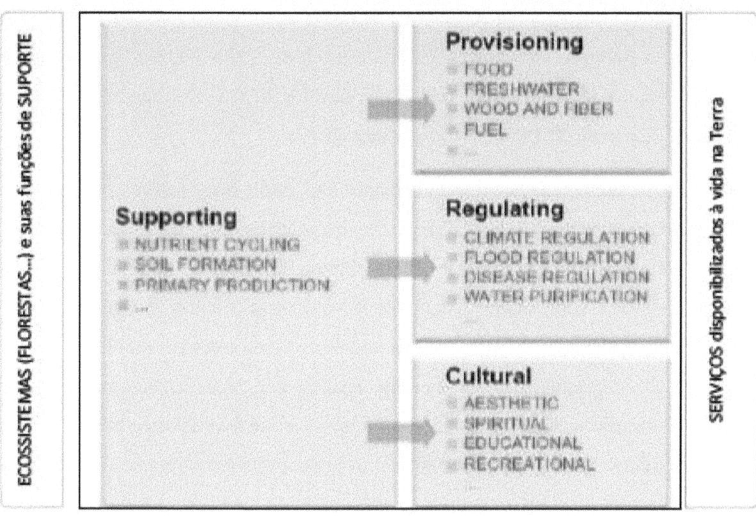

Figure 1: *Services provided by ecosystems for human well-being* (Source: Adap. from WHO, 2005, p. 15).

According to (WHO, 2005, p. 27) the structure and functioning of the world's ecosystems has changed rapidly, especially in the second half of the 20th century, causing profound changes in the diversity of life on earth:

- Extensive land has been converted into agricultural land, with the result that today a quarter of the earth's surface is covered by agricultural land;

- In the last decades of the 20th century, approximately 20% of the world's coral reefs were lost and another 20% degraded;

- The withdrawal of water from rivers and lakes has doubled since 1960, the vast majority (70%) of which is used for agriculture;

- Since 1960, nitrogen flows have doubled and phosphorus flows have tripled;

- Since 1750, the atmospheric concentration of carbon dioxide has increased by around 32% (from 280 ppm to 376 ppm in 2003).

(Muacahila, 2017) states that in Angola the degradation of ecosystems has caused great damage to the country. The impact of the destruction of the value of environmental services in Angola is estimated at 6.7%, representing a loss of around 517.4 billion US$/year (Sutton et al., 2016, cit. by Muacahila, 2017).

Recent scientific studies suggest that, on the one hand, tropical and subtropical forests act as a means of transporting humidity, influencing the global circulation system that affects the transportation of harmful agents and precipitation itself; on the other hand, they can help achieve several of the many Sustainable Development Goals of the United Nations 2030 Agenda (SDGs) (FAO, 2018). In fact, forests and their trees are sources of energy, as they provide vegetable fuel for cooking, heating and industrial needs, as well as protecting the watersheds that make it possible to generate hydroelectric power, among other contributions. That's why it's important to protect this valuable resource, since, as (Cunha, 2011) states, the environmental dysfunctions of human activities are always the result of economic intervention that ignores the depletion of natural resources.

The following figures show the current reality, in terms of a) proportions of wood use as fuel in the various regions and countries of the world (Figure 3) and b) percentages of households using forest biomass as fuel for cooking and sterilizing water in the various countries (Figure 4).

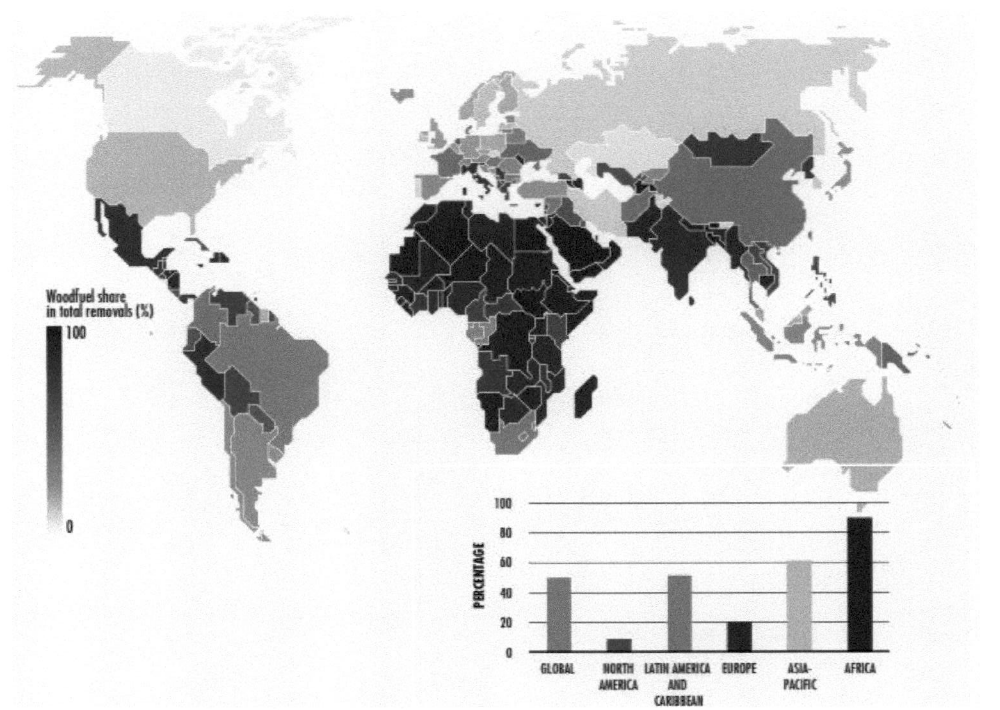

Figure 2: **Proportions of wood used as fuel in the various regions and countries of the world** (Source: FAO, 2018, p. 35).

17

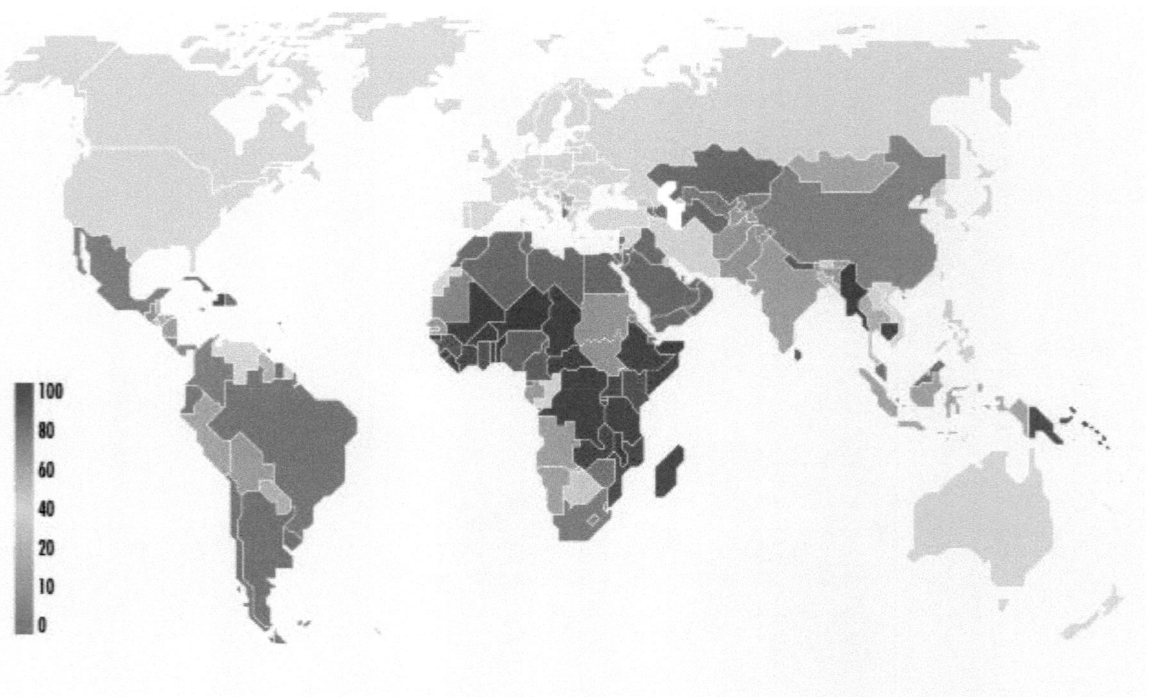

*Figure 3: **Percentage of households by country that depend on vegetable fuel** (Source: FAO, 2018, p. 33).*

The figures associated with these inventories show that at least around 88.5 million people in Europe and North America make wood their main source of heating (FAO, 2018). In the particular case of Angola, Figure 4 shows how around 60% of the population uses vegetable fuel to meet their energy needs (heating, cooking).

Given this reality, all actions that lead to the preservation and conservation of all types of forests (stopping deforestation, restoring degraded forests, increasing reforestation, afforestation), especially the implementation of their sustainable management, are necessary in order to maintain the levels of benefits that forests provide (Figure 2). This must also be the way forward to combat other harmful practices that devastate forests, such as burning and the indiscriminate felling of trees for various purposes. For, according to (Hinrichs and Kleinbach, 2009), the lifespan of a resource depends very much on the speed with which it is used, whether sustainably or unsustainably. Figure 5 clearly illustrates the current global situation of forest fires. In global terms, we can see that Angola does not stand out in a negative way, with a burnt area of only between 1000-2000 km^2 (which represents 12% of the national territory if we consider a burnt area of 1500 km^2), in the period 2003-2012, but which should still worry us.

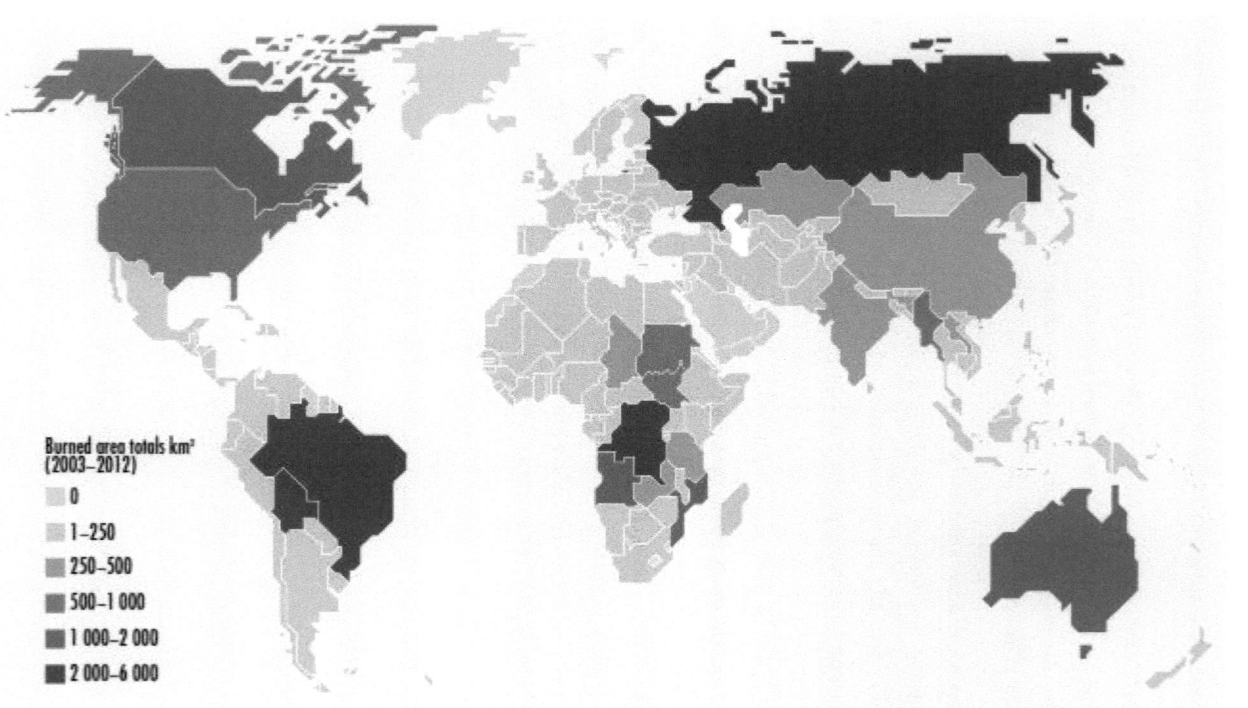

Figure 4: **Area of forest burned between 2003 and 2012** (Source: FAO, 2018, p. 55).

The reality of forest fires not only affects countries' performance in the expected sustainable management of forests, but also jeopardizes the achievement of the commitments made in this area in the context of fulfilling the SDGs. In Angola, where data on this phenomenon is scarce, it is difficult to predict whether SDG 15 (Protect terrestrial life) will be achieved.

According to (Earth Alliance et al., 2007) tropical deforestation has implications for climate change, as it is responsible for the emission of around 2 billion tons of carbon per year into the atmosphere, and has also contributed to its decline at rates of around 5% per decade. Table 2 shows how the activity that contributes to the deforestation of tropical forests contributes to global greenhouse gas emissions of around 27%.

Table 2 **Contribution of deforestation to global greenhouse gas emissions** *(Source: Houghton, 2005, cit. by Aliança da Terra et al., 2007, p. 1).*

Type of gas	Contribution to the increase in the Greenhouse Effect (%)	Annual Emissions from Deforestation	Deforestation's contribution to total global emissions (%)	Contribution of deforestation to the global increase in the greenhouse effect (%)
CO_2	58	2.2 Pg C	26	15
CH_4	21	275 Tg CH_4	48	10
$N O_2$	6	5.4 Tg N O_2	33	2
Total	85			27

The figures in Table 2 not only show how important it is to reduce and/or eliminate actions that contribute to deforestation, thus ensuring a reduction in greenhouse gas emissions into the atmosphere, but also call for governments to create targeted strategies to reduce or eliminate emissions.

1.3 Angola and the Commitment to Sustainable Forest Management

1.3.1 Forest Management Instruments in Angola: the Forestry Regulation

Angola has its own legislation to intervene in the management of its forests, which is complemented by international standards and other instruments. The various legal instruments available include the Constitution, the Basic Law on the Environment, the Land Law, the Basic Law on Forests and Wildlife, the Spatial Planning Law and the Forestry Regulations. Without prejudice to the interaction and subsidy that must exist in the articulation and application of the various legal instruments listed, forest management in Angola finds its support in Presidential Decree No. 46/14, of February 25, which approves the National Action Program to Combat Desertification, Presidential Decree No. 171/18, of June 25 (Forestry Regulation), which regulates Law No. 6/17, of January 24 (Bases of Forests and Wildlife).

The Forestry Regulation stands out as the main instrument for sustainable forest management in Angola. In general terms:

- Regulates the sustainable management of forest resources and the ecosystems themselves; and

- It establishes rules for the conservation and rational use of forest resources, taking into account the environmental, social, economic and cultural dimensions of these resources.

In terms of guiding principles, without prejudice to compliance with those contained in other legislation also applicable in this sector, the regulation takes into account the following principles selected by us, as they seem to be most closely linked to our concern under analysis, deforestation:

- "Social justice, well-being and citizen participation". It ensures the participation of citizens and takes into account their concerns and needs in the forest management process.

- "Ecological balance". Certainly to ensure a balance between the supply and demand of resources, thus ensuring the protection and preservation of natural resources for current and future generations.

- "Prevention and precaution". In situations that indicate damage to the environment and in which prevention fails, even if there is no conclusive scientific evidence, apply the precautionary principle: *in dúbio pro defendant* (in this case the environment, the forest).

- "Integrated management of forest resources". Management should result in compatibility between forest management, land use planning, river basin management and the protection of the general environment.

- "International cooperation". This consists of coordinating and creating bilateral instruments to manage common forests with other countries (see the case of the Maiombe forest in Cabinda and the integrated forest in the Okavango region in Cuando Cubango).

According to Article 5 of the Forestry Regulation, the national forestry heritage is divided into two groups: 1) all populated and non-populated forest land, provided it is classified as such, and 2) natural and planted forests. And according to its potential, geographical location and form of use, the forest estate is classified into the following types of forest (no. 2, article 5):

a) *Protective forests* - made up of plant formations whose function is protection and conservation, maintenance and regeneration, and which are under a special management regime.

b) *Special purpose forests* - made up of plant formations with national defense, permanent protection, conservation, scientific research, landscape protection, leisure and cultural functions.

c) *Production forests* - these are natural or planted forests located outside conservation areas and made up of plant formations with high economic potential intended for commercial exploitation.

Production forests (the case of the local Munhino forest) are more vulnerable to the actions that cause their degradation, as they are outside protection and conservation areas. Their resources are authorized by law to serve economic purposes aimed at: a) the subsistence of local communities, b) cutting timber and harvesting other products for their own consumption, and c) the general exploitation of their products for various purposes (DecPre, 171/18, article 8). In fact, Article 21 of the Regulation grants free use rights for subsistence and community enjoyment, without being subject to prior authorization or respecting the vegetative rest period. These are the purposes recognized for communities:

1. Food, energy, medicinal, therapeutic and cultural;

2. Obtaining forest raw materials for construction, rural furniture and handicrafts produced by community residents;

3. Logging and deforestation for subsistence farming.

The Forestry Regulation, in addition to leaving forest management in the hands of local communities, makes an exception and gives the right to commercial exploitation of forest products in forests that have sufficient forest resource potential, but with authorization from the authorities (no. 3, art. 21).

1.3.1.1 Quantities to be harvested and trade in forest products in the communities

In this regulation, the right to subsistence use and community enjoyment of natural forest resources, although governed by the customary norms and practices of the peoples of the communities, must also comply with certain restrictions:

- *Quantities to be extracted*. The extraction of firewood, building stakes and other woody material for subsistence purposes and community enjoyment should not exceed an average of 3 m^3 or equivalent per person per year. Here the important question, which perhaps constitutes a loophole, is who should control these quantities and how? Since an unjustified excess gives rise to the payment of an exploitation tax (art. 23).

- *Production of firewood and charcoal for commerce*. The large-scale production of wood fuels (firewood and charcoal) must be exploited in forest plantations established for this purpose (art. 24).

- *Marketing of forest products by communities*.

> - In the communities, the products harvested in the forest are intended for consumption, and must not circulate outside these communities or be marketed (art. 25);

> - Sales of wood fuel, building materials and non-timber products are allowed only between neighbors in the communities (art. 26);

- Outside the communities, the sale of handicraft items is permitted, in quantities not exceeding two sacks of charcoal and one stere of firewood, stakes, poles, or equivalent building material (art. 27).

- *Monitoring intra- and extra-Community marketing.* This task is entrusted to "supervisory agents in general", "community observers in particular", and "in conjunction with the traditional authority of the area" (art. 28).

1.3.2 The Commitment to Sustainable Forest Management

1) Protect, restore and promote the sustainable use of terrestrial ecosystems, 2) sustainably manage forests, 3) combat desertification, 4) halt and reverse soil degradation and 5) halt biodiversity loss are the 5 major tasks set out in SDG 15 (Protect terrestrial life), which is part of the 17 goals of the United Nations 2030 Agenda.

To meet this challenge, Angola has defined the following 8 national priorities (INE, 2018):

1- Drawing up strategies and implementing actions to adapt to and mitigate climate change, particularly measures to combat drought and desertification;

2- Implement nature and biodiversity conservation actions and strengthen sectoral policies related to the protection of wild flora and fauna;

3- Strengthen waste collection and sorting actions, promote environmental awareness and education actions and environmental monitoring;

4- Prevent natural hazards and protect populations in vulnerable areas;

5- Ensuring effective territorial management of conservation areas;

6- Manage the country's forests based on the principles of sustainability;

7- Ensuring sustainability in forest management; and

8- Promote the reforestation of degraded areas.

SDG 15 (Protecting Life on Land) has 12 targets and 14 indicators, of which Target 15.2 stands out for this study, stating "By 2020, promote the implementation of sustainable management of all types of forests, prevent deforestation, restore degraded forests and substantially increase afforestation and reforestation efforts globally". This target is monitored by Indicator 15.2.1, defined as "Progress towards sustainable forest management". It can be seen that this indicator does not have a "quantity value" available which represents the measurement baseline for assessing the final result. However, the sub-indicators defined in Table 3 (1- rate of net change in forest area, 2- above-ground forest biomass reserve, 3- proportion of forest area located in established protected areas, 4- proportion of forest area under a long-term management plan, and 5- proportion of forest area certified but under independent forest management) can serve to measure progress towards sustainable forest management in this particular case.

Table 3 shows data from an initial assessment of progress in sustainable forest management in the world's various sub-regions, the results of which suggest that there is still a long way to go and major challenges to overcome in this area, including for the Sub-Saharan region (which includes Angola), indicating the danger of not meeting the targets set.

Table 3Regional progress in achieving sustainable forest management for each SDG 15 indicator (Source: Adap. from FAO, 2018, p. 61).

World Region	(1) Net area of modified forest	(2) Above-ground forest biomass reserve	(3) Proportion of forest area located in legally protected areas	(4) Proportion of forest area in long-term forest management plans	(5) Proportion of certified forest area under independent forest management
World	●	●	●	●	●
North America	●	●	●	●	●
Europe		●	●	●	●
Latin America and the Caribbean	●	●	●	●	●
Central Asia	●	●	●	●	●
South Asia	●	●	●	●	●
East Asia	●	●	●	●	●
Southeast Asia	●	●	●	●	●
West Asia	●	●	●	●	●
North Africa	●	●	●	●	●

Sub-Saharan Africa	●	●	●	●	●
Australia and New Zealand	●	●	●	●	●
Other parts of Oceania	●	●	●	●	●

Caption:
● None/Small change Positive change Negative change Non-certified area ●

With regard to Sub-Saharan Africa, where Angola is located, the data in Table 3 shows how the effort made so far to implement sustainable management of its forests, especially to respond to the following national priorities of the eight defined (6- Manage the country's forests based on the principles of sustainability; 7- Ensure sustainability in forest management; and 8- Promote reforestation of degraded areas), has only managed to achieve progress in the indicators (3) Proportion of forest area located in legally protected areas and (4) Proportion of forest area in long-term forest management plans. The data also shows that the country and the region in which it is located have failed to comply or have complied poorly with the indicators (1) Net area of modified forest and (2) Above-ground forest biomass reserve. There has also been no progress on indicator (5) Proportion of certified forest area under independent forest management. In fact, it is in this last indicator that most of the country's forestry activity falls under and is reflected, and this is the case of the forest in the Munhino territorial area, which, as it is a *natural production forest,* should be given special attention in terms of reforestation, sustainable management and more effective monitoring, thus guaranteeing its sustainability.

In this regard, (MADR-MUA, n.d.) states that there is no coordinated information system on renewable natural resources in the country, nor are there well-founded management plans for the exploitation of forest resources, with some information based on estimates, of limited coverage and of dubious value. Naturally, there are many weaknesses at this level. There is also the fact that forest management is still centralized and does not favor the involvement of partners with interests in the sector, thus compromising feasible decision-making and effective monitoring of activities.

As a result of these dysfunctions, every day we see massive actions without environmental impact studies, such as anarchic burning, deforestation for the practice of shifting agriculture, the devastation of large areas with trees for the agribusiness industry and mining, the establishment of precarious human settlements, overgrazing, the uncontrolled production of firewood and charcoal, the extraction of building materials, among others, thus leading to a reduction in forest area and deforestation on the one hand, and desertification on the other.

1.4 Current overview of Angola's forest landscape: strategies to reduce deforestation and degradation

Addressing the current state of Angola's forests has not been an easy task for most people. This is due to the fact that information and data on this sector is not accessible to the general public, and information, as in many other sectors and national services, is not published or if published is not made available. However, with a bit of luck, some media outlets manage to retrieve some information, data and concerns about the forestry sector in Angola.

Quoting the Director General of the IDF, Mr. Tomás Caetano, Agência Angola Press was able to report that the first inventory of the national forest showed that Angola had 60 million hectares of forest by 2017, and that the annual deforestation rate was 8.2%. For the National Director, this rate is acceptable, as it allows the available forest resources to be planned and managed smoothly, as "the aim is to guarantee the sustainability of the forest exploitation process, controlling the way in which industries and populations exploit the available forest cover" (ANGOP, 2017). For his part, the National Director of Biodiversity, Mr. Nascimento António, speaking to Jornal de Angola on World Forest and Tree Day, warned that charcoal production damages hectares of forest (JN, 2019).

These public speeches by senior managers and executives linked to the institutions responsible for forestry give us an idea of the scale of the major problem our forests are suffering from - the lack of sustainable management.

To help find solutions that can change this situation, in Angola and other parts of the world, some scholars have advocated specific strategies that can be adapted to each situation to reduce deforestation and forest degradation, especially that resulting from illegal actions/activities. It should be noted that these strategies should be aligned with the prevention and control actions that official bodies (IDF, economic police and local administrative bodies) are already applying on the ground against these illegal activities.

Of the various strategies that can be adopted and considering the dynamics of the systemic approach of the Forest Management Cycle (Figure 6), we highlight these two (Aliança da Terra et al., 2007) that can be adapted to our reality:

- Increasing the transparency and dissemination of existing information and data on deforestation, making it known to government bodies and non-governmental and civil society organizations concerned with this issue; and

- Implement and develop mechanisms to mobilize will and financial funds aimed at the conservation of forests and the sustainable use of their resources.

However, the Angolan government's long-term strategy for combating deforestation is defined in the National Action Program to Combat Desertification (PANCOD)[2]. Forests fall under the heading of "Preservation, conservation and sustainable management of natural resources", which defines the strategy, objectives, actions and targets (Table 4).

As you can see, this strategy and the actions designed can be combined with the creation of new protected areas and the expansion of existing protected areas, the creation of legal tools adapted to each local reality to combat infringements, as well as putting these laws into practice to punish offenders and thus reduce the feeling of impunity, among other measures.

Essentially, the implementation of this national strategy to combat deforestation and desertification of the national territory must be based on a systemic and dynamic approach made up of the following stages: Environmental Licensing, Monitoring, Inspection, Legal Accountability (Figure 6).

Table 4Strategy and actions defined to combat deforestation and desertification in Angola (Presidential Decree no. 46/14, of February 25).

Strategy: to provide a framework for intervention and formulation of public and private projects that promote an increase in vegetation cover, availability of vegetation resources and knowledge of native vegetation for planning use and development.		
Actions	Objectives	Goals
- Introduction of alternative sources of domestic energy production (improved stoves, gas stoves) in areas with fragile ecosystems.	- Managing traditional systems of charcoal and	- Improved management of natural forest resources and plantations.

Actions	Objectives	Goals
- Introducing modern forms of coal production and increasing yields. - Training artisans in modern charcoal-making techniques and improved stoves. - Inventory of forest formations producing firewood and charcoal.	firewood production, distribution and consumption	- Clarification of the functions of forest ecosystems.
- Planning the use of land, water and forest resources. - Inventorying the country's floristic potential and drawing up red lists of vulnerable and endangered species. - Drafting appropriate legislation and regulatory standards governing the use and utilization of forest resources.	- Improving the management of forest resources, particularly in transhumance areas.	- Monitoring herd numbers. - Cultivation of pastures, control and regularization of transhumance and marketing of surplus livestock. - Up-to-date knowledge of the country's forestry potential and its categorization.
- Introduction and implementation of management plans for natural forest formations. - Promoting legal measures that encourage the full utilization and use of waste resulting from the forestry production chain. - Taking advantage of existing forest plantations, prioritizing their exploitation and renewal, as well as promoting new plantations. - Promoting agro-forestry practices.	- Promoting the sustainable management of natural productive forests and natural plantations	- Substantial reduction in pressure on native forest. - Increase in the supply of wood for various purposes and transfer from places of abundance to places of shortage. - Implementation of forests with priority given to sensitive areas and those susceptible to natural disasters. - Increase in the plant *continuum* and availability of plant material for various human purposes.

Actions	Objectives	Goals
- Drawing up sensitivity and soil erosion charts. - Promoting forest replanting for soil conservation and recovery of degraded forest areas. - Creation of forest polygons for productive purposes, particularly for the supply of wood, firewood and charcoal. - Creation of environmental education programs/projects for the population on measures to protect, preserve and sustainably use flora resources.	- Combating land degradation and desertification.	- Mobilizing communities for environmental protection and conservation actions. - Renovation and installation of forest nurseries in all provinces and localities lacking native vegetation for energy and pasture. - Knowledge of sensitive areas and control of ravines and desertification.

- Creation of community forest plantations to produce wood. - Educating the population on measures for the protection, conservation and sustainable use of the resources available to them.		- Controlling pressure on forest resources and increasing their availability. - Prevention and mitigation of the effects of human activities on the environment. - Increased awareness of life's dependence on basic natural resources.

*Figure 5: **Systemic approach in the Forest Management Cycle** (Adap. from Aliança da Terra et al., 2007, p. 5).*

Each element of this systemic approach has the following meaning: *environmental licensing* (process of authorizing the exploitation of forest resources by the communities, prescribing the ways in which the activities must be processed and what quantities must be subtracted per year), *monitoring* (verification task assigned to the communities themselves to periodically check that the organization of their activity and the procedures being used in the exploitation of natural resources are correct), *inspection* (an activity that allows inspection authorities such as the IDF to see if the exploitation is being carried out by the communities correctly and if it complies with the purpose assigned to it in the license) and *legal accountability* (the act of punishing offenders by the legal authority).

1.5 Conclusion of the chapter

The literature review made it possible to understand the various aspects of forest management, especially the practices that contribute to deforestation. This exercise revealed that the forest in the Munhino region, in the national context, falls into the category of community forest, a forest with a production function. According to the literature, intervention actions in these forests should result in an improvement in the quality of the exploitation of their resources. To this end, a systemic approach based on key elements such as *environmental licensing, monitoring, supervision* and *legal accountability* must be taken into account so that management is sustainable and does not compromise future generations.

CHAPTER II
METHODOLOGICAL ASPECTS AND STUDY AREA

Chapter 2. METHODOLOGICAL ASPECTS AND STUDY AREA

2.1 Methodology, procedures and study area

This chapter characterizes the study area, as well as explaining how the research was conducted, following the parameters for collecting and processing the information. In the meantime, the work carried out followed stages and procedures, the overview of which is shown in Figure 6 below.

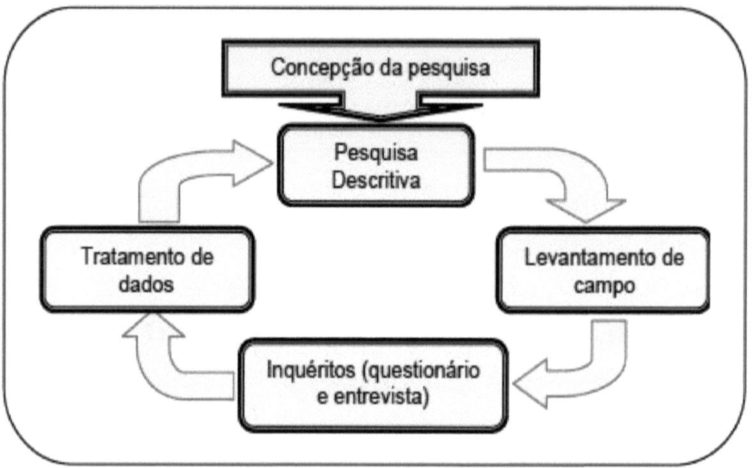

*Figure 6: **Stages and procedures of the research.***

2.1.1 Characterization and location of the study area

Climatic aspects and location

Bibala is one of the five municipalities of Namibe province, about 160 km from the provincial capital (Moçâmedes) and located in the far northeast of the province. It is made up of five communes: Bibala, Caitó, Lola, Kapangombe and Munhino, the last of which is the study area - the Munhino commune (Figure 7).

Figure 1 **Location of the study area**. *Source: Urso, 2013*

In the province of Namibe, the most significant vegetation cover is located in the northern interior, in the municipalities of Bibala and Camucuio, even though there are no species that are of major interest to the timber industry. The municipality of Bibala, with its mountainous terrain, has a climate classified as semi-arid (Tornthwaite scale), dry steppe (Koppen scale), or dry semi-arid (Rivas-Martinez scale) (Muacahila, 2017). Its average daily temperatures vary between 16ºC and 28ºC and it experiences average annual rainfall of between 23 mm and 144 mm (Figure 8).

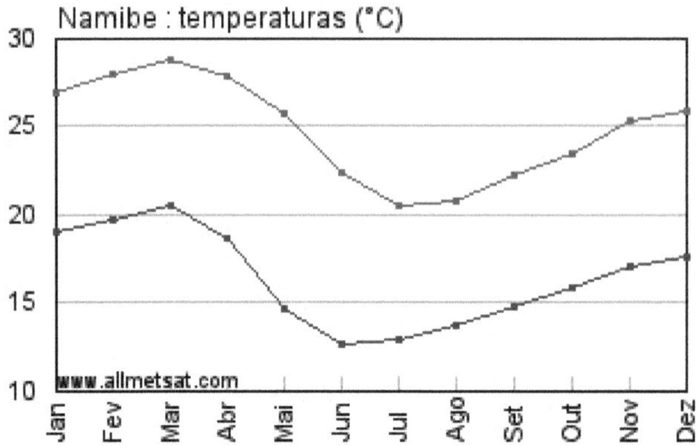

Figure 2Average temperatures and rainfall in the municipality of Bibala (Source: www.meteoblue.com).

As the municipality is in the hot and dry transition zone, its dominant soils are *lithosols and* rocky *terrain* (poorly developed soils with stony material) and *oxisialites* (rich in clay and iron oxide minerals). As far as phytography is concerned, the *shrub steppe of the sub-desert belt* predominates, with outcrops of mutiati forests and savannah herbaceous communities. The predominant shrub species used by the population for various purposes are *Colophospermum monpane* (mutiati) and *Spirostachys africana* (omupapa) (Diniz, 1991). Overall, the Bibala region has a very low tree density, totaling 508 individuals in the sampling units studied and representing 169 trees per hectare (Joaquim, 2018). The rainy season generally runs from October to April, with prolonged interregnums in January (meteoblue.com; Diniz, 1991). The municipality is criss-crossed by intermittent dry rivers (rios de areia or mulolas), most notably the Munhino river, after which the Munhino commune is named. The waters in the Munhino region are moderately saline, which makes them suitable for human consumption, watering livestock and irrigating crops in the lower areas (GPN, 2007).

Economic aspects

Until 2007, the land structure of the municipality of Bibala was 223,757 ha, distributed as follows: farming (123,142 ha), livestock (99,670 ha) and agriculture (945 ha) (GPN, 2007). As far as agricultural activity is concerned, as the municipality is in a hot-dry transition zone, the main crops are grown on dryland farms and small irrigated farms. This is a favorable area for crops such as sugar cane, beans, sesame, massango, massambala and fruit trees, as well as the production of castor beans, sisal and tobacco. As an area with natural pasture resources, it has potential for pastoralism, beef cattle farming, *caracul* sheep and goat farming, as well as for transhumance at certain times (Diniz, 1991). The activity rate in the municipality of Bibala (58.6%) exceeds that of the provincial seat (Moçâmedes, 53.5%) and also the provincial average (50.2%). In Namibe province, the municipalities of Bibala and Moçâmedes have the lowest unemployment rate (14%), below the provincial average (18%) (INE, 2016). In Bibala, even if not very significant, it is common to find activities linked to wholesale and retail trade, and other informal activities that support families.

Demographic aspects

With a surface area of 8,534.844 km^2 and 64,504 inhabitants, the municipality of Bibala has a population density of 7.6 inhabitants/km^2 , making it the second most inhabited municipality in the province, behind the municipality of Moçâmedes (292,535 inhabitants). The average household size in Bibala is 5 members. The commune of Munhino (the study area) has an area of 4,043 km^2 and a population of between 10,000 and 20,000 inhabitants. Bibala's population is very young, with individuals aged 0-14 accounting for 51.3% and the 65 and over age group accounting for 4%. The illiteracy rate in this municipality is around 34.2% and only 0.3% of the population has completed higher education (INE, 2016). The images in Annexes 3A, 3B, 3C and 3D illustrate some of the reality of the area studied.

The following images show some socio-environmental aspects of the current reality in the Munhino communities (Figure 9: A, B, C, D, E and F).

*Figure 3*Aspects of anthropogenic actions exerted on the Munhino forest and its current state of deforestation.

Methodology and procedures

If methodology is the application of procedures and techniques aimed at building knowledge about a given reality or subject, the method represents the path, the way of approaching the procedures and techniques applied (Pradonov and Freitas, 2013; Carmo and Ferreira, 1998). A descriptive and qualitative investigation was carried out, as it describes the real situation of forest use by the communities of the Munhino locality, as well as the meaning that people give to their daily lives. In order to better understand the situation and the problem under study, documentary analysis was used (official documents, printed sources, legislation and regulations) and questionnaire surveys were applied (to IDF-Namibe technicians and officials) and interviews (to the local population living in various communities).

The following equipment/materials and software were used to acquire field information and process the data: Garmin *etrex* GPS (which was used to survey the geographical coordinates for mapping the location), QGIS 3.4 version LTR (Madeira) and Google Earth Pro (which enabled the map of the study area to be produced), Excell software from the Microsoft Office 2013 Student package (which enabled the graphs to be constructed), and IBM SPSS Statistics 12 software (which used the Chi-square test (x^2) of independence to test the association between variables in the sample, as described in (Marôco, 2014; Laureano, 2013). In all analyses, a probability of $\alpha = 0.05$ was considered and Monte Carlo Simulation was used whenever the conditions for approximating the test distribution to the Chi-square distribution were not met.

2.1.3 Sample and sampling procedure

When carrying out the surveys, not knowing the exact number of people living in the Munhino commune (due to a lack of official data) made it impossible to opt for parametric or probabilistic sampling methods, so we opted for non-parametric or non-probabilistic methods.

Thus, the *convenience sampling method* was used to obtain the sample for interviewing the individuals from the four communities, which made it possible to easily approach the 42 people from the community who were available and others by accident, in a cheap and quick way (Marôco, 2014; Hill and Hill, 2012). The interviews (Annex 1 and Figure 10) were carried out with the help of four (4) people (one of whom acted as an interpreter when communicating with the locals), at three different times (in the last week of January/2020 and during the month of February/2020) at the four sampling points/locations (P1, P2, P3 and P4) (see Figure 7).

As for the technicians and managers surveyed, the so-called *expert sampling method* was applied, as they are people with knowledge of the sector under study and the problems that affect it (Marôco, 2014). Ten questionnaires (Annex 2) were distributed in January 2020, so only three questionnaires were returned, which represents a return rate of 30%.

Figure 4*Interview conducted in one of the villages in the Munhino communities*.

2.2 Conclusion of the chapter

The Munhino region, like the municipality in which it is located, is in a hot-dry transition zone, where the main crops are grown on rainfed farms and small irrigated farms. This is a favorable area for growing various crops. This region has a very young population represented by individuals aged 0-14 (51.3%), with a high illiteracy rate (34.2%). Knowledge of the region's characteristics and dynamics helped us to carry out a descriptive and qualitative investigation, in order to describe as reliably as possible the real situation of forest use by the local communities and to better understand the meaning that people give to their daily lives in the communities of this region.

CHAPTER III
ANALYSIS AND DISCUSSION OF RESULTS

Chapter 3. ANALYSIS AND DISCUSSION OF RESULTS

3.1. Organization of the analysis and discussion of the issues

This chapter presents and discusses the results of the surveys addressed to the sample studied and to IDF technicians and managers. Before the actual analysis and discussion, the sociodemographic characterization of the interviewees is presented. The analysis of the results follows the sequence of the questions in the surveys, taking into account the relevance of the questions. The questions considered are identified with the letter Q followed by the question number Q7 (question seven of the survey).

3.2 Sociodemographic characterization of the sample studied

The 42 individuals interviewed were 22 (52.4%) male and 20 (47.6%) female, with ages ranging from under 18 (n= 3) to 46 to 60 (n= 3). The most representative age groups were 26 to 35 (n= 14; 35%), 18 to 25 (n= 11; 28%) and 36 to 45 (n= 9; 23%). Most of the respondents (n= 35; 83%) live in the catchment area of Munhino. The respondents' level of education is very low, with the majority being illiterate (n= 32; 76%). In terms of occupation, the groups that stand out are those who are self-employed (n= 19; 45%) and those who are unemployed (n= 15; 36%) (Figure 11).

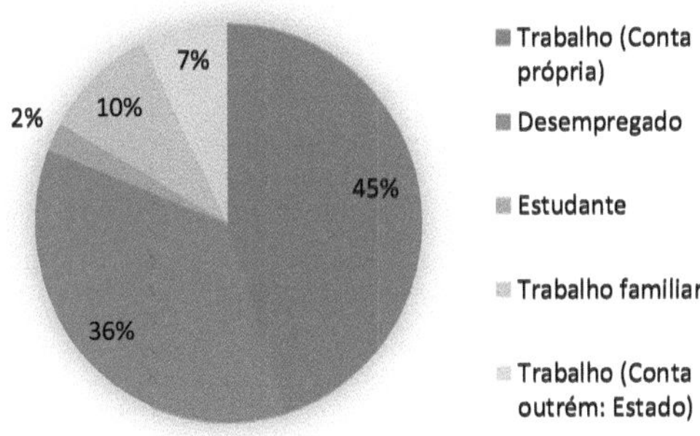

Figure 5Interviewees' working conditions.

3.3 Analysis of interviews with community members

Q6- If you do any forestry work (logging, charcoal production, ...) and for how long. When asked if they have any occupation related to forestry, the majority of interviewees said yes (n= 31; 74%), while the rest said no (n= 9; 21%) or didn't answer (n= 2; 5%) (Figure 12).

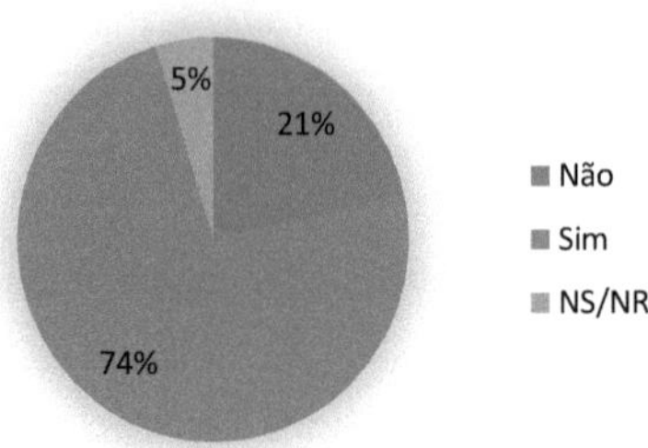

Figure 6**_Forestry activity._**

A large number of people who exploit forest resources, namely timber extraction for various purposes and charcoal production, have been doing so for at least 5 years (n= 14; 33%), while the second largest group of "exploiters" can't remember how long they've been doing it (n= 11; 26%). This strong connection between people and the local forest, in terms of exploiting its resources, makes sense insofar as, as we saw above, the majority of these people are self-employed or unemployed, so exploiting forest products is the form of work/employment and the source of their family income. Figure 13 shows how most people have been exploiting and living off forest resources for more than five or 10 years.

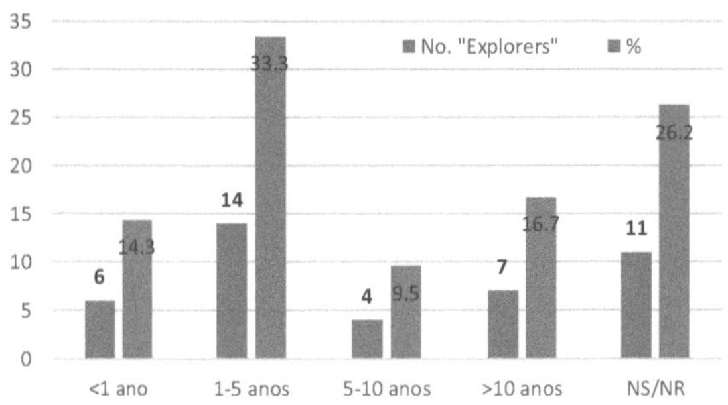

Figure 7Time spent exploiting forest resources.

Q7- **On the purpose of exploiting forest resources**. The forest products extracted from the local forest are used for three purposes: 1) the majority use these products for their own consumption and for trade (n= 20; 47%), 2) others sell the products produced (n= 15; 36%), and 3) still others use these products for their own consumption (n= 7; 17%). The data therefore shows how the local forest represents a source of employment and income for many families (Figure 14). In this case, the Chi-square test of independence showed that there is a relationship between the variables place of residence and the purpose for which forest products are produced ($X^2_{(6)}$ = 19.680; p = 0.005). This means that the residence factor has an influence on the type of purpose given to the products produced: residents direct their production essentially to two purposes (self-consumption and trade), while non-residents direct their production to trade[3] .

---

[3] See test results in Annex 5.

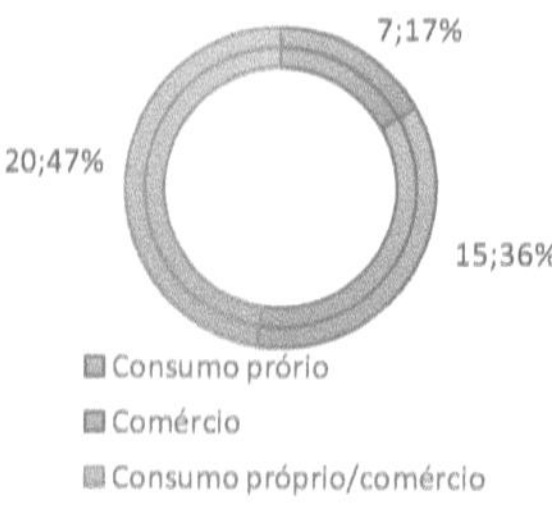

Figure 8Purpose given to products extracted from the forest by the local population.

Q(8 to 10)- **About hunting and cattle-raising practices in local communities**. Leaving aside the extraction and sale of timber and charcoal, as the majority of interviewees (n=35; 83%) confirmed (see Figure 13), the members of these communities also engage in other activities such as agriculture (n= 20; 48%), cattle breeding (n= 12; 29%) and hunting (n= 2; 5%) (Figure 15). It's easy to see the effort made by families to practise agriculture and the fact that the majority of families took up this activity at least five years ago. Raising their own livestock is also beginning to gain ground (n=12; 29%).

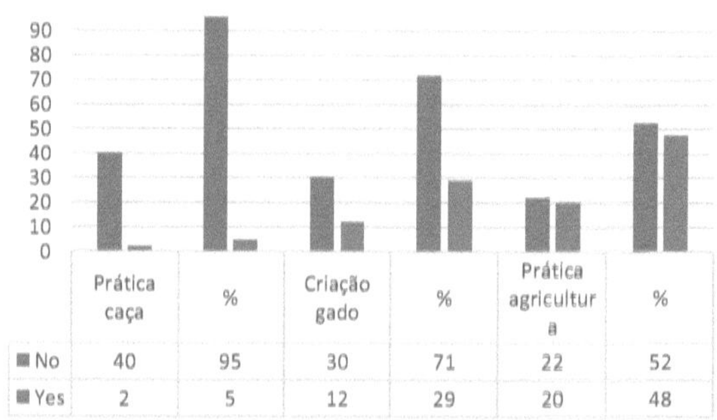

	Prática caça	%	Criação gado	%	Prática agricultura	%
No	40	95	30	71	22	52
Yes	2	5	12	29	20	48

Figure 9Hunting and raising cattle in the communities.

Q(11) - **Main problems facing the local forest in the opinion of the interviewees**. The five main problems or dangers that the local forest currently faces in the opinion of the interviewees and according to Figure 16 are:

1. Poor legal framework and inter-institutional cooperation (n= 39; 93%);
2. Deforestation and unregulated charcoal production (n= 26; 60%);
3. Lack of protection and preservation by the local community (n= 22; 52%);
4. Lack of enforcement and criminalization of offenders (n= 19; 45%);
5. Abandonment of traditional Environmental Management (n= 13; 32%).

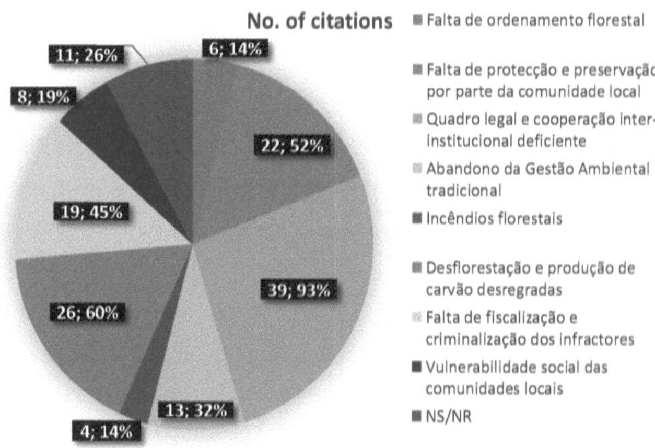

Figure 10Main problems and dangers facing the local forest in the opinion of the interviewees.

These figures, while on the one hand showing that the population has some idea of the problems facing their forest and caused by their actions, on the other hand show that around a third of those surveyed have no idea or are unaware of the problems facing their forest (n= 11; 26%). In fact, this data points to the need to raise awareness among the members of these communities about adopting new ways of living with their forest and the need for the legal authorities to organize this activity and monitor it, and with the help of the communities to find the best ways of intervening in the local forest (Figure 17).

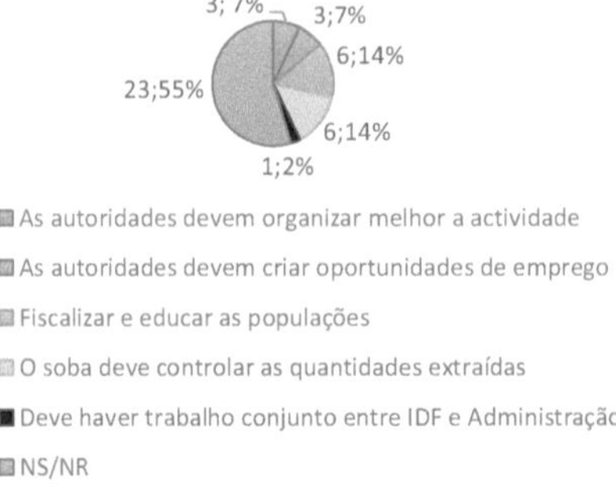

*Figure 11**Measures suggested by interviewees to improve local forest management*.

<u>Questions about the future of the forest and its importance for local communities</u>

Q(12) - **Functions that the local forest performs in the opinion of the interviewees**. For the interviewees from the Munhino communities, the local forest performs important functions (Figure 18) such as a) "Biodiversity conservation" (n= 28; 67%), b) "Climate regulation" (n= 13; 31%), c) "Timber extraction" (n= 9; 21%), d) "Charcoal production" (n= 8; 19%) and others. However, it can be seen that many do not know or do not appreciate the importance that the local forest plays (n= 18; 43%) and the need for its preservation and conservation, in order to continue to sustain community life today and in the foreseeable future. This once again highlights the need to raise awareness and educate people about the right way to relate to this community asset - the local forest. In fact, it is this message and need that people have (see suggested measures in Figure 17).

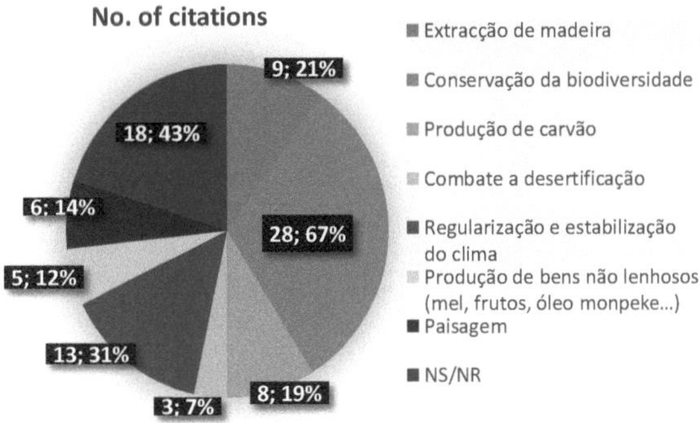

Figure 12Functions performed by the local forest in the opinion of the interviewees.

Q(13)- **Willingness of interviewees to participate in the search for solutions for the preservation and conservation of the local forest**. With regard to this question, the figures show that the interviewees have different positions (Figure 19). On the one hand, there are those who are willing to participate and contribute to defending their forest (n= 15; 36%), and on the other hand, those who don't know or don't have an idea. There were more cases of willingness in the age groups 36 to 45 years old (n= 6), 18 to 25 years old (n= 3) and 26 to 35 years old (n= 3), compared to those who are not willing or who don't know or don't have an opinion. However, the Chi-square test of independence showed that there was no relationship between the variables age and willingness to participate in the joint effort to find solutions to the problems facing the community ($X^2_{(10)}$ = 14.272; p = 0.134). On the contrary, the gender variable is related to the feeling of willingness to participate in any effort aimed at solving the current problems facing the local forest ($X^2_{(9)}$ = 44.734; p = 0.001)[4].

---

[4] See the results of the two tests in Annexes 3 and 5.

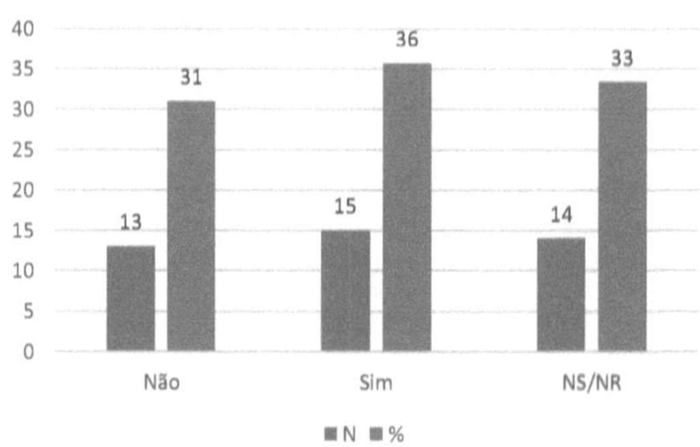

Figure 13Opinion of the interviewees about their willingness to find solutions for the local forest.

Q(14) - On whether the interviewee had ever taken part in a forest awareness session. As shown in Figure 20, the majority of interviewees (n= 37; 70%) have never taken part in any training, clarification or awareness-raising session about the forest, its importance in the lives of communities and the need for it to be looked after. This may partly explain the aggressive way in which communities exploit the local forest, living only for today and tomorrow belongs to God (as they say in popular parlance). Of the five interviewees who said they had taken part in a workshop of this nature, 3 (three) had taken part in events held by student groups (from Bibala, Lubango and ISCED), 1 (one) had taken part in a meeting held by a group of national and foreign engineers, and another had taken part in a workshop held by the NGO COSPE.

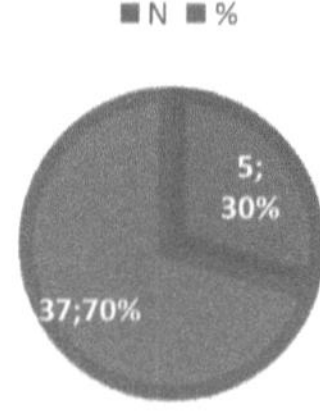

Figure 14Participation in awareness-raising sessions and meetings.

Q(15) - **Forecast of the state of evolution of the local forest**. On this question, the figures in Table 5 show that the majority of interviewees have some awareness of how their current actions contribute to the degradation of their forest, highlighting aspects such as "mutiati population", "wood and charcoal production" and "undergrowth". Another relevant aspect is the fact that many of these people don't know or don't have an opinion, which once again reinforces the need to intensify awareness-raising and socio-environmental education in these communities.

Table 5Forecast of the evolution of the local forest in the view of the interviewees.

Aspects considered	Increase	%	Keep	%	Decrease	%	NS/NR	%
Forestry in general			2	5	2	5	19	45
Agricultural area			12	29	6	14	24	57
Population of Mutiati			3	7	26	62	13	31
Hunting reserve			2	5	19	45	21	50
Wood and charcoal production			1	2	27	64	14	33
Landscape search for tourism			0	0	15	36	27	64
Matos			0	0	27	64	15	36

3.4 Analysis of questionnaires addressed to IDF technicians and managers

Questions about human practices in the forest and their purpose

The results of the three questionnaires answered by some of the IDF technicians/managers (out of a planned ten) show that these services are aware of the existence of forest product extraction practices, and that the main purpose of this activity is commerce (n= 2; 67%), followed by "own consumption and commerce" (n= 1; 33%). It can also be seen that in forest management, specifically the control and monitoring of illegal activities practiced by communities, the services have applied regulations and other legal instruments. However, this situation can be counteracted to some extent:

- For lack of evidence or practical information on the ground;

- In the opinion of the people interviewed, who said there was a lack of "supervision and criminalization of offenders" (n= 19; 45%; Figure 16), which led them to suggest actions such as "supervising and educating people" and that "the authorities should better organize the activity" (n= 14; 35%; see Figure 17).

Questions about the IDF's action in the management and organization of the local forest and future vision

These technicians/responsible people who were questioned testified that some events had taken place which involved local communities, in particular the *workshop* organized by the NGO COSPE. This was confirmed by some of the local people interviewed. Respondents confirm that the IDF is facing a number of difficulties in carrying out its work properly and point to the main problems facing the local forest:

1- Lack of protection and preservation by the community;
2- Legal framework and poor inter-institutional cooperation;
3- Deforestation and unregulated coal production;
4- Lack of supervision and criminalization of offenders.

To overcome these problems, respondents point to a number of measures:

1- Fight poverty by improving the living conditions of communities;
2- Encouraging agro-horticulture;
3- Reforestation; and
4- A ban on the felling of trees to produce charcoal for commerce.

3.5. Some of the most significant results of the Chi-square test of independence

To help interpret some of the less visible aspects discussed here, specifically the relationship between some of the variables in the survey aimed at the population of the Munhino community involved in the study and engaged in the exploitation of timber resources (timber and charcoal production), Table 6 below shows the statistically significant relationships obtained using the chi-square tests of independence.

Table 6Most significant incidences found using the Chi-squared test (X^2) of independence.

Question number and content	Q3 Sex	Q7 What do you use the wood and charcoal produced for?	Q13 Would you be willing to take part in working sessions to find solutions to protect the forest?
Q1 Whether you live in Munhino or not.		Since most of the respondents live in the catchment area of Munhino (n= 35; 83%), they are the ones who channel the forest products produced for their own consumption and trade.	
Q2 Age			It is in the most representative age groups in the communities, 18 to 25 years old (n= 11; 28%), 26 to 35 years old (n= 14; 35%), and 36 to 45 years old (n= 9; 23%), where there is more willingness and availability to contribute to the search for solutions.
Q13 Would you be willing to participate in working sessions to find solutions to protect the forest?	There is equal willingness among men and women to participate in finding viable solutions for their communities.		

3.6 Conclusion of the chapter

This chapter made it possible to discuss the results of the interview surveys conducted with the population studied (a sample of 42 individuals) and the results of the questionnaires applied to IDF technicians and managers. The interaction established shows how these communities recognize the problems facing the forest. From the lack of protection and preservation by the community, to the lack of supervision and criminalization of the offenders who devastate it with unbridled actions such as cutting down trees and producing charcoal for trade. It is also interesting to see that the same populations are willing to participate in the search for solutions to improve their living conditions, with actions such as the promotion of agro-horticulture, reforestation and a ban on the felling of trees to produce charcoal for commerce.

CHAPTER IV

CONTRIBUTIONS TO SUSTAINABLE FOREST MANAGEMENT IN THE MUNHINO COMMUNITY

Chapter 4. CONTRIBUTIONS TO SUSTAINABLE FOREST MANAGEMENT IN THE MUNHINO COMMUNITY

4.1 Proposed Actions/Activities for Sustainable Forest Management in the Munhino Communities

In this chapter, based on the problems diagnosed and which lead to unsustainable management of the local forest, we present our contribution by proposing actions/activities and recommendations that we believe will be useful in dealing with the current constraints and problems affecting the natural resource Forest in this area of national territory. The actions/activities presented (Table 7), together with the desired objectives, the expected benefits and those involved in the desired actions, are not miracle solutions, nor are they the only ones, but if applied within local constraints, they could contribute to a better relationship between communities and their environment (forest), as well as reducing the aggressiveness with which the forest resource is affected and consequently resulting in better forest management and community well-being, on the one hand. On the other hand, these actions proposed here are in line with the concerns raised by the respondents, both the community population and the technicians/people in charge.

Table 7Proposed Actions/Activities for Sustainable Forest Management in the Munhino Communities.

Action / Activity	Objective(s)	Benefits	Actors / Participants
1- Drawing up regional, provincial and local plan(s)/program(s) containing instructions for sustainable forest management appropriate to each territorial reality	- Inventory the existing biomass in each territory; - Regulate the annual dynamics of forest resource exploitation, in terms of allocation and quantity; - Promoting the rational use of biomass energy	- Control of available forest resources; - Sustainable forest management; - Improving the dynamics of production, distribution and consumption of forest products.	- IDF; - Provincial government(s); - Municipal administrations; - Communal Administrations; - Higher Education Institutions; - Community representatives.
2- "Preparation" of a Local Forestry Management Plan (PGEFL), in accordance with Article 85 of the Forestry Regulations.	- Encouraging exploitation under the Forest Exploration Regime (collective and individual); - Indicate appropriate harvesting techniques and the annual quantities of forest resources to be extracted	- (Encouraging) selective exploitation; - Maintenance and renewal of tree species; - Reduced proliferation of operators; - Greater availability of forest resources over the long term; - Creation of organized and controlled employment; - Improving the well-being of communities.	- IDF; - Bibala Municipal Administration; - Munhino Communal Administration; - Higher Education Institutions; - Community representatives.
3- Creation of mechanisms to involve traditional authorities and local communities in the decision-making process regarding forest management.	- Integrating local socio-economic and environmental needs into official plans and programs; - Commit local populations and other stakeholders to the cause of preserving and conserving common natural resources	- Communities identify with the actions and rules outlined and practice them; - Communities defend and protect their natural asset (forest)	- IDF - Other official authorities - Stakeholders

(Continued from Table 7_1)

Action / Activity	Objective(s)	Benefits	Actors / Participants

4- A ban on the production of charcoal for commerce and the promotion of the extraction of other potentialities of the local forest (beekeeping, monpeke oil, ...).	- Drastically reduce forest deforestation; - Encourage the practice of other activities offered by the forest; - Valuing the potential of the local forest.	- Maintenance of the local forest without major modifications; - Diversification of forest exploitation; - Production of more and varied forest products for the communities; - Prosperity of communities.	- IDF; - Bibala Municipal Administration; - Munhino Communal Administration; - Community representatives.
5- *Workshops* and other actions in the community	- Raising people's awareness of good deeds; - Providing socio-environmental education to communities for better management of forest resources; - Teaching better forest resource management practices; - Disseminating knowledge about good environmental management practices.	- Acquisition of good and better knowledge about the sustainable management of natural resources; - The evolution of communities in terms of their relationship with their surroundings; - Better environment and quality of life in communities.	- IDF; - Bibala Municipal Administration; - Munhino Communal Administration; - NGOs; - Environmental associations; - Educational institutions; - Other stakeholders
6- Recruiting "monitoring assistants" from the local communities (sort of like forest rangers)	- Record in real time the activity carried out in each community related to the extraction of forest resources; - Inhibit deforestation of the forest; - Informing the competent authorities of dangerous developments in forestry activity.	- Control and preservation of the local forest; - Job opportunities in the community;	- IDF; - Bibala Municipal Administration; - Munhino Communal Administration
7- Afforestation and reforestation for identified purposes.	- Producing raw materials for charcoal production; - Recovering deforested and degraded areas; - Protecting the environment and the local forest; - Create plantations for energy purposes and community trade.	- Maintenance of the local forest without major modifications; - Diversification of forest exploitation; - Production of fuel (firewood and charcoal) for own consumption and trade;	- IDF; - Bibala Municipal Administration; - Munhino Communal Administration; - Community representatives.

		- Improved community and family income; - Job creation.	

4.2 Conclusion of the chapter

The approach taken in this chapter, based on the problems diagnosed, which lead to unsustainable management of the local forest, has helped to make contributions through the proposed actions/activities and recommendations set out in Table 7. In short, the actions/activities presented, together with the desired objectives, the expected benefits and the actors to be involved in the desired actions, if applied appropriately, can contribute to a better relationship between communities and their environment (forest) and thus reduce the aggressiveness with which the forest resource is acted upon.

CONCLUSIONS

CONCLUSIONS

Finally, it is important to note that the study carried out showed how important it is to reduce or eliminate actions that contribute to deforestation. In this case under study, as it is a community forest classified as production forest, in order for the proposed intervention actions to result in an improvement in the quality of the exploitation of forest resources by local communities, the systemic approach suggested in the literature must be taken into account, always bearing in mind its key elements: environmental licensing (the process of authorizing the exploitation of forest resources by communities, prescribing the ways in which activities should be processed and what quantities should be subtracted per year), monitoring (the verification task assigned to the communities themselves to periodically check the organization of their activity and the procedures being used in the exploitation of natural resources are correct), inspection (an activity that allows inspection authorities such as the IDF to see if the exploitation is being carried out by the communities correctly and if it complies with the purpose assigned in the license) and legal accountability (which is the act of punishing offenders by the legal authority).

In this study it becomes clear how the communities of Munhino face basic socio-environmental problems that contribute to the current state of degradation of the local forest, among others:

- The lack of job opportunities;

- The poverty that exists in communities;

- The high level of illiteracy;

- The precarious living conditions;

- The lack of awareness-raising and socio-environmental education for the population on aspects related to the exploitation of natural resources;

- The lack of active and preventive monitoring of illegal activities carried out by local populations, which cause incalculable damage to the local forest; and

- The failure to welcome and take into account the position of communities in government and/or administrative processes that influence their living conditions.

The people in the local communities interviewed, although the majority are illiterate (76%), are nevertheless aware of how the lack of supervision and punishment of their actions by the competent authorities contributes to the intensification of inappropriate practices in the exploitation of the forest resources available in their communities. The members of these communities show a willingness to participate in building solutions to resolve the current problems affecting their communities and degrading their living conditions. Another relevant aspect is the recognition of the need to implement intensive awareness-raising and socio-environmental education actions within these communities, without prejudice to the other actions proposed, given that there is a low level of education, which sometimes means that people do not reflect on the socio-environmental and economic problems in which they are immersed.

Fulfilling the fifth objective of this work ("Designing a sustainable intervention proposal"), seven actions/activities have been indicated for intervention in these communities (Table 7), the expected results of which will make it possible, in an initial phase, to minimize the impacts of current anthropic actions on the local forest; and in a later phase and over time, to consolidate the actions and results of the initial phase, in a joint effort between official entities, local communities and environmental associations, NGOs and other interested parties.

The perception of the motivations that lead to the way in which local forest resources are exploited and the problems arising from this unsustainable form of exploitation diagnosed here, the education and schooling of the local population should be the first line of intervention. Because only with educated citizens can the desired results be achieved.

In short, the planned actions contribute to sustainable forest management and, consequently, to saving the forest. In this sense, it is necessary to take into account the local socio-environmental problems diagnosed and apply the required actions in an appropriate manner in order to achieve good results and guarantee a significant influence on raising awareness and making local communities responsible for reducing deforestation and consequently deforestation.

BIBLIOGRAPHICAL REFERENCES

_ANGOP (2017). "Angola has 60 million hectares of forests". Online news from Agência Angola Press, February 13, 2017 (available at: www.m.portalangop.co.ao; access:26/04/2020).

Earth Alliance et al. (2007). Three Key Strategies for Reducing Deforestation (available at: www.ipam.org.br; access: 24/03/2020).

Carmo, H. and Ferreira, M.M. (1998). Metodologia da investigação: Guia para auto-apredizagem. Lisbon: Universidade Aberta.

Carapeto, C. (2004). Fundamentals of ecology. Lisbon: Universidade Aberta.

Clemente, M.H. (2020). Environment Source of Life: Reasons for the existence of life on Earth. Vol. 1. São Paulo: Metabooks.

_Cunha, F.M. (Coord.).(2011). Environmental Management and Sustainability. Lisbon: Verlag Dashöfer.

_DecPre (46/14). National Action Program to Combat Desertification (PANCOD) (Presidential Decree no. 46/14, of February 25.

_DecPre (171/18). Forestry Regulations (Presidential Decree no. 171/18, of July 23).

Diniz, A.C. (1991). Angola: the physical environment and agricultural potential. Lisbon: Institute for Economic Cooperation.

FAO (2006). Global Forest Resources Assessment 2005. Rome. Italy.

FAO (2007). Definitional issues related to reducing emissions from deforestation in developing countries. Forests and Climate Change Working Paper 5. Rome.

FAO (2018). The State of the World's Forests 2018 - forest pathways to sustainable development. Rome.

Faucheux, S. and Noel, J-F. (1995). Economics of natural resources and the environment. Lisbon: Instituto Piaget.

GPN (2007). Namibe Province Integrated Development Plan. Sectoral Studies. Volumes 3 and 4.1. Namibe.

Hill, M.M. and Hill, A. (2012). Research by questionnaire. Lisbon: Edições Sílabos.

Hinrichs, R.A. and Kleinbach, M. (2009). Energy and the Environment (3rd ed.). São Paulo: Cengage Learning.

_INE (2016). General population and housing census - 2014. Definitive results Namibe province. Luanda: Reprographics Division.

_INE (2018). Sustainable Development Goals. Report on the Baseline Indicators. Agenda 2030. Luanda: INE-Departamento de Informação e Difusão.

IPCC (2003). Definitions and methodological options to inventory emissions from direct human-induced degradation of forests and deforestation of other vegetation types. Framework for a definition. Japan: Institute for Global Environmental Strategies.

ITTO (2005). Revised ITTO criteria and indicators for the sustainable management of tropical forests including reporting format. ITTO Policy Development Series No. 15.

Joaquim, P.C. (2018). Exploitation of Colophospermum mopane (mutiaty) and the conservation paradigm of natural ecosystems in the province of Namibe - Angola. Master's dissertation in Management and Conservation of Natural Resources at the University of Évora.

_JN (2019). "Deforestation worries the Ministry of the Environment". Online news item from Jornal de Angola, March 26, 2019 (available at: www.jornaldeangola.sapo.ao; access:26/04/2020).

_Laureano, R.M.S. (2013). Hypothesis testing with SPSS: My quick reference manual (2nd ed.) (revised and updated). Lisbon: Edições Sílabos.

MADR-MUA (undated). Forestry policy, wildlife and conservation areas. Discussion document (2nd version). (available at: www.fao.org; access: 11/03/2020).

Marôco, J. (2014). Statistical analysis with asPassa Statistics. 6th ed. Lisbon: ReportNumber.

Motta, R.S. (2007). Environmental economics. Rio de Janeiro: FGV.

Muacahila, A.N.S. (2017). Sustainable development strategies for the dry regions of southern Angola - Curoca River Basin. Thesis presented to fulfill the requirements for a PhD in Environmental Sciences and Engineering. Department of Environment and Planning, University of Aveiro.

Pradonov, C.C. and Freitas, E.C. (2013). Metodologia do trabalho científico: Métodos e técnicas da pesquisa e do trabalho académico (2nd ed.) (E-book). Novo Hambuego-Rio Grande do Sul (Brazil): Universidade FEEVALE.

WHO (2005). Ecosystems and human well-being: health synthesis. France: Millennium Ecosystem Assessment (available at: www.millenniumecosystemassessment.org; access: 30/3/2020).

_www.meteoblue.com (access: 24/02/2020)

yes
I want morebooks!

Buy your books fast and straightforward online - at one of world's fastest growing online book stores! Environmentally sound due to Print-on-Demand technologies.

Buy your books online at
www.morebooks.shop

Kaufen Sie Ihre Bücher schnell und unkompliziert online – auf einer der am schnellsten wachsenden Buchhandelsplattformen weltweit! Dank Print-On-Demand umwelt- und ressourcenschonend produzi ert.

Bücher schneller online kaufen
www.morebooks.shop

info@omniscriptum.com
www.omniscriptum.com

Printed by Books on Demand GmbH, Norderstedt / Germany